SCSC Newsletter Volume 31 No. 1
February 2023

SCSC Publication Number: SCSC-182

Contents

FOR EVERYONE WORKING IN SYSTEM SAFETY

www.scsc.uk/sss

SSS' 23

Details of this and
other events at:
www.scsc.uk

Safety-Critical Systems Club Annual Symposium

www.scsc.uk

THE SAFETY-CRITICAL SYSTEMS CLUB

31st Safety-Critical Systems Symposium

7-9th February 2023, in-person in York, UK

The Safety-Critical Systems Symposium in 2023 (SSS'23) will be held in the historic city of York, UK. This event comprises three days of presented papers, including keynote presentations and submitted papers. There will also be a banquet on the Wednesday evening with after-dinner speaker.

This event will be run in-person at the Principal York hotel (located next door to York station).

The Symposium is for all of those in the field of systems safety including engineers, managers, consultants, students, researchers and regulators. It offers wide-ranging coverage of current safety topics, focussed on industrial experience.

It includes recent developments in the field and progress reports from the SCSC Working Groups. It takes a cross-sector approach and includes the aerospace, automotive, defence, health, marine, nuclear and rail areas.

The symposium features are:

- Seven keynote presentations and talks on submitted papers;
- Updates from the SCSC Working Groups;
- 'Room 101' entertainment with audience participation and submitted questions
- A banquet with after-dinner speaker on the Wednesday;
- Proceedings book and working group guidance books;
- Visit and demonstrations at the University of York Institute of Safe Autonomy;
- Special beer (or non-alcoholic alternative) designed and commissioned for the symposium;
- Safety Activities;
- Social events in York.

The symposium is a regular event and fosters a community spirit in the field: it is a great place to network, learn about the latest practice in safety and develop new business contacts.

For **further information, exhibition** and **booking queries** please contact: Alex King, Dept of Computer Science, University of York, Deramore Lane, York, YO10 5GH. Email: alex.king@scsc.uk

For **technical aspects** and **talks, speakers, abstracts, papers** or **posters**, please contact: mike.parsons@scsc.uk

Editorial

Last year I saw the retirement of a couple of individuals in my company who had pretty much been around for as long as I could remember, certainly in excess of 30 years. Seeing them go was a bit of a shock; like suddenly finding your favourite armchair missing from the corner of the room, and it made me reflect on the general state of the industry. The following figure shows the demographic distribution from the attendees of the last Safety-Critical Systems Club Symposium in February 2022.

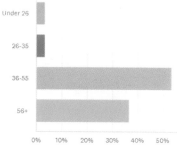

With a sample of around 125, I hope it not too much of a leap to suggest that this might give a reasonable reflection of the industry, and it's a sobering picture. Given that many pension funds allow retirement at 55, in theory at least, almost 40% of the workforce could disappear tomorrow.

At SSS'22 we reflected on the past but also looked to the future and while we were excited about the technological challenges ahead, I think one of our most important challenges is to ensure that we will have the people in sufficient numbers with the right skills and competencies required to meet these challenges.

So where do we go from here? I believe one of the most important initiatives is the SCSC's Safety Future Initiative (SFI) that sets out to support the development of the next generation of safety practitioners. I would therefore whole-heartedly encourage you all to help support this. The SFI are currently looking for volunteers and I'm sure we can all play a part: from offering to act as a mentor through to looking at the junior people in your own companies – is there anyone you could encourage to develop an interest in systems safety? Are there other experienced people who could cross-train into the discipline?

To see the work of the SFI first hand, Laure Buysse has very kindly provided an article to introduce herself and provide her thoughts on her career and the work of the SFI.

There are also other interesting feature articles, with Nick Hales providing an example of using his Layered Enterprise Data Safety Model to gain insights into the Manchester Arena bombing of 2017. Dhanabal Arunachalam discusses a proposed safety analysis technique that allows synergy between safety and cybersecurity during the development of autonomous vehicles and Divya and Martin Atkins introduce a new tool to help those implementing the Data Safety Guidance to manage data safety risks.

We have an event report by Bernard Twomey from "The Future of Testing for Safety-Critical Systems" seminar and Dave Banham provides a report of the latest SCSC Tech Trip to the RAF Museum Midlands (Cosford). As introduced in the last edition, in our "Recent Publications" section we have another book review with Malcolm Jones providing his thoughts on the 2nd Edition of the book "System Safety for the 21st Century" by Richard A. Stephans.

Our 60 second interview is with Phil Koopman an internationally recognised expert on Autonomous Vehicles.

Paul Hampton
SCSC Newsletter Editor
paul.hampton@scsc.uk

In Brief

Final report on Boeing 737 MAX crash sparks dispute over pilot error

The long-delayed Ethiopian government report into the crash of Ethiopian Airlines Flight 302 that killed all 157 people on board laid blame solely on Boeing.

The subsequent French and American critiques — a rare fracture among the safety authorities participating in an air accident investigation — don't dispute Boeing's role, but present a fuller picture of the tragedy's cause. *seattletimes.com*

B-2 nuclear bomber fleet grounded amid search for safety defects

All 20 of the US Air Force's B-2 Spirit bombers are grounded as the service hunts for potential safety

defects, a spokesperson confirmed.

The Air Force paused the fleet's operations after a bomber was damaged on 10th Dec at Whiteman Air Force Base, Missouri. An undisclosed in-flight malfunction forced the crew to make an emergency landing on Whiteman's runway, where firefighters extinguished flames at the scene. *airforcetimes.com*

GM's Cruise robotaxi unit faces safety probe after three accidents

The National Highway Traffic Safety Administration (NHTSA) is investigating three crashes involving driverless taxis operated by General Motors's Cruise, just as the autonomous-vehicle division is poised to expand service.

Officials are looking into Cruise vehicles that braked suddenly – resulting in three crashes in which human motorists rear-ended the robotaxi – or cars that unexpectedly pulled over and stopped, according to a NHTSA. *spokesman.com*

People who skipped their COVID vaccine are at higher risk of traffic accidents, according to a new study

During the summer of 2021, Canadian researchers examined the encrypted government-held records of more than 11 million adults, 16% of whom hadn't received the COVID vaccine. They found that the unvaccinated people were 72% more likely to be involved in a severe traffic crash.

The authors theorise that people who resist public health recommendations might also "neglect basic road safety guidelines." *fortune.com*

Applying the Layered Enterprise Data Safety Model

The Glade of Light memorial opened in 2022 to commemorate the victims of the Manchester Arena bombing of 2017, one of the deadliest terrorist attacks in the UK. Nick Hales examines how the Layered Enterprise Data Safety Model (LEDSM) and supportive techniques can provide insights into how to prevent the worst aspects of an attack of this nature and in managing the sharing of information in its aftermath.

The Manchester Arena suicide bombing was an example of how badly emergency services safety communication can fail the public, despite the fact that the multi-agency integrated emergency management process is statutory under the Civil Contingencies Act 2004 and is referred to as a Local Resilience Forum. The report from the Public Enquiry highlighted that duties of the responders include *"to cooperate; to share information; to assess risks in their area; to plan for emergencies"*. Unfortunately, it is clear the emergency services failed in these four areas because they assumed:

- the cooperating and sharing of information had been thought through in planning
- cooperating and sharing information would be simple
- risks had been assessed by someone and mitigated
- the granularity of analyses would enable effective action and all would know what to do

It is for those reasons that the incident has been chosen to illustrate how the LEDSM [1] process can effectively prepare the communication network for emergencies. The importance of communication is illustrated by examining the Joint Emergency Service Interoperability Principles, (JESIP), which feature in the inquiry report and are listed here with annotations by the author:

- Co-location (to improve communication)
- Communication (that is what this article is about)
- Coordination (facilitated by effective communication)
- Joint Understanding of Risk (one big risk being poor communication)
- Shared Situational Awareness (facilitated by reliable and available communication networks)

Brief Recap on LEDSM

- The term LEDSM is used because not all important, independent nodes of a network are necessarily organisations. Just like businesses, Enterprises can be one person or many, and this is reflected in the many to one and one to many relationships involved in LEDSM planning. The flexibility and granularity of LEDSM diagrams means that communication from a multi-level LEDSM can connect with one person who owns their own single level Enterprise.
- In electronics, a seven layer model has been used to establish most communication between machines, the Open Systems Interconnection (OSI). It is considered that a seven layer model at maximum for any enterprise in a network should be adequate for emergency communication planning. Although somewhat arbitrary, it is considered that anything larger than the number of levels complex electronics manages with, would become cumbersome and begin therefore to make successful communication less likely.
- The process of LEDSM analysis including associated techniques is illustrated in the figure below (See [1] for further information and diagrams)

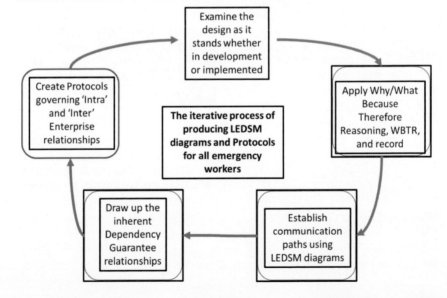

Detail relevant to the Manchester Arena concert bombing is given later in the text. Any Enterprise may have many links from various layers but the use of LEDSM diagram enables appropriate granularity so that those connections to Enterprises not involved in the problem being addressed may be omitted, even down to individuals being offered their own LEDSM diagram with just the connections they need to make shown.

Identification of Partner Organisations

This article does not go in depth into the identification of partner Enterprises though that is an important part of de-risking any safety-critical event or process as that is a choice of the organisations involved. Here the Enterprises that were involved are simply shown below in a LEDSM maximum configuration diagram, (i.e. with seven layers that not each Enterprise may eventually use but is always a useful starting point).

The lower levels are equivalent to the electronic checking levels in the OSI model. They represent those operatives who may or may not currently be involved in the creation of Protocols, suggestions for other Enterprises to be involved, the development of Dependency-Guarantee relationships, or links to higher and lower ranking persons that full analysis of potential emergency events may prove to be potentially beneficial.

The links shown are simply those between the Gold or in LEDSM terms Enterprise level and the Resilience Forum as complex relationships at a top level abstraction are not possible unless on very large sheets of paper.

The convention of referring to the top three levels as Gold, Silver and Bronze, as used by the Manchester Resilience Forum is used throughout to ensure the Inquiry Report terminology and this paper are easily understood.

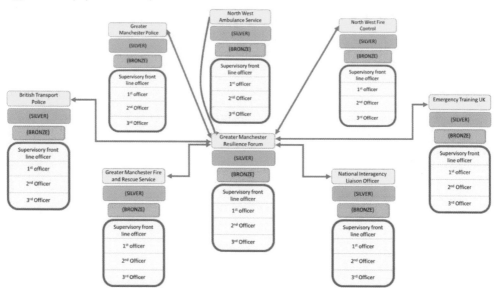

Timeline of Events

Having established the primary Enterprises involved in an Emergency in Manchester, those involved would move to the next stage and then go through the cyclical process, as illustrated in the first figure.

To show the effectiveness of LEDSM, and how it would reduce risk, a timeline of events is used which is given in the following table. It shows a few of the problems in the first ten minutes of the incident and comments indicating more detail of the failures. Only a subset of the full plethora of problems that occurred in the golden hour immediately after the bomb, in which lives can generally be saved, can practically be reproduced in this article. Indeed, even the problems of first ten minutes, as given by the Inquiry Report [2] had to be reduced to some of the most significant.

TIME	ACTION	COMMENT
22.31	Bomb goes off	
22.32	NWAS should have contacted their Hazardous Area Response Team (HART)	It took until 22.39 for Tactical Commander Annemarie Rooney to tell the GMP HART to employ their specialist skills at the bombing site.
22.34	GMP control duty officer Inspector Sexton	Takes Gold and Silver GMP roles as well as both Tactical and Strategic Firearms Commander roles
22.35	Inspector Sexton should have declared a "MAJOR INCIDENT"	Total failure of training. As Gold Commander he must take total responsibility for this.
22.36	NWFC control should have mobilised GMFRS	In the absence of LEDSM protocols, NWFC supervisor prevents mobilisation insisting that a National Interagency Liaison Officer (NILO) must be consulted first.
22.36	BTP officers at the site of the bomb fail to declare a MAJOR INCIDENT and fail to send a METHANE message	LEDSM protocols carried with officers detailing their expected behaviour would have ensured the required messages were passed on.
22.36	NWAS control tries to Contact GMP but has to wait in a queue for 2 minues	LEDSM analysis would have shown a reserved radio channel was an essential fall back measure for lack of telephone communication as would multiple telephone options including land-line and mobiles.
22.36	GMP Inspector Michael Smith declared a Rendezvous Point (RVP) to GMP control	In the event the message was only passed to NWFC in reasonable time 22.40 but other organisations did not get it or got it an hour later, (GMFRS at 23.54).
22.37	ETUK personnel did not pass on a METHANE message to NWAS as ther training instructed them to do.	The LEDSM analyses would have identified this as a protocol and a Dependency-Guarantee relationship established that would and should be a high priority and wold have been know to be such by ETUK staff
22.37	Incident log at BTP records inability to contact GMP	LEDSM analysis would have shown the need for multiple channels of communication between the Enterprises.

What/Why Because Therefore Reasoning, (WBTR)

LEDSM applied to any system, whether it is human only, human and machine or machine to machine communication, is firstly about one loss, the 'loss of correct communication'. 'Correct' being used here as a term that encompasses timeliness, availability and accuracy among other qualitative terminology. Secondly, it is about capturing concerns visually. Hence, in the WBTR stage, conversations may take place such as the hypothetical one given here in dialogue involving a Front Line Officer, (FLO) and a BRONZE Commander, (BC):

FLO:- If I am first at the scene WHAT do I or my team members need to do to inform others of the scene?

BC:- One of you will need to send a METHANE message. The question is WHAT could go wrong?

FLO:- I might get injured or killed.

BC:- OK, then we need all your team trained to send the METHANE message.

These conversations can then be gathered into formal decisions as to what must happen, as below.

WHY/WHAT	BECAUSE	THEREFORE
What to do if first at scene	METHANE messages are priority	Need appropriate comms and training
What if dedicated METHANE officer cannot send	There may be hazards that injure or at worst kill the officer	Other trained and equipped personnel needed
Why might the METHANE message not get through	Jammed lines or problem broadcasts like interference	Extra diverse communication needed
Why does everyone need two channels	All need to know the METHANE message and one channel may fail	Diverse comms needed for all involved
What Enterprises need METHANE messages	All Resilience Forum members need METHANE data	Have group phone number and a common radio channel
What if mobile and radio channels are down	The METHANE message and others agreed are essential	Declare METHANE as text message and move position until it is sent by the mobile

When WBTR is used, one can start with a picture of an envisaged system or one may have to start with something only understood in verbal terms at the outset. This is why the Manchester Arena bombing is a good example to describe here as it is only one of many scenarios the emergency services may get caught up in, e.g. a plane crash as at Lockerbie, or a pub blast, none of which can be pictured particularly well in the initial analysis because of the variety of locations at which an emergency can occur.

As can be seen, the problems in communication should not have occurred if this formal methodology had been used as they are plain to see and moreover, in the next sections the development of the LEDSM diagram can be shown to produce an easily understood (and adjustable if necessary) picture of who should be contacting which role in which Enterprise.

The Initial Internal LEDSM Block

A meeting will need to be called at this point for the first iteration of the cyclic LEDSM process introduced earlier. This meeting is best facilitated by Suitably Qualified and Experienced Personnel, (SQEP). It would be only a matter of guessing what partner stakeholder Enterprises require and may supply in terms of data. Therefore, it is important to establish a LEDSM Block of one's Enterprise to provide an opening gambit for the subsequent meeting with other stakeholder Enterprises, members of the Resilience Forum in this case.

Some negotiation of positions on necessary LEDSM levels and who communicates to whom will be required internally, prior to the negotiations on positions in other Enterprises. The fundamental of this phase, is to establish the protocols that define the data that will be passed up and down internally in the Enterprise, the intra-communication as shown in next figure that shows the (up to 7) levels in an Enterprise between each of which protocols will exist.

These protocols will define information from the Managerial level downwards (dashed), and personnel offering suggestions upwards to change their protocol (dotted). The top three levels may be considered equivalent to the Gold, Silver and Bronze levels in the current command and control structures.

In the case used here, British Transport Police (BTP) are to contact Greater Manchester Police (GMP) with a METHANE message, which stands for:

- **Ma**jor Incident?
- **E**xact location?
- **T**ype of incident?
- **H**azards?
- **A**ccessibility?
- **N**umbers of casualties?
- **E**mergency services attending?

The relevant instruction among all the protocol instructions from the GOLD Enterprise level downward in this Inter-Enterprise protocol creation phase, would be simply to "Complete a METHANE message and send to all other Enterprises requiring it".

This is passed down through SILVER and BRONZE levels who may add to the protocol created at the GOLD level for additional requirements they need of the staff assigned to deal with the incident.

Enterprise (GOLD)	The top three layers are equivalent to the Gold, Silver and Bronze levels currently used
Organisational Unit (SILVER)	
Optimising (BRONZE)	
Supervisory front line officer	These four layers are the Front Line personnel. Their behaviours should be as automatic as possible, normally through rehearsal and adherence to agreed protocols.
1st officer	
2nd Officer	
3rd Officer	

Also at this stage, initial Dependency Guarantee Relationships (DGRs) need to be proposed. One such could be for instance, a BTP officer, who may get there first, stating that he or she will require a guarantee that a message will be received, (a dependency), "*A Rendezvous Point notification within 4 minutes of arrival at the scene*" and guarantee to "*Send a METHANE message within 3 minutes of arrival at the scene*".

That officer has then created both a "Dependency Relationship" and a "Guarantee Relationship" for future negotiations with other Enterprises which the senior personnel in the officer's Enterprise will be aware of and agreed to.

A first meeting with partner Enterprises

The next figure shows an example of the finer detail for the links between GMP and BTP, which relate to the event shown at 22.36 in Timeline of Events table. At that point in time and subsequently, a METHANE message should have been sent but was not in the actual incident.

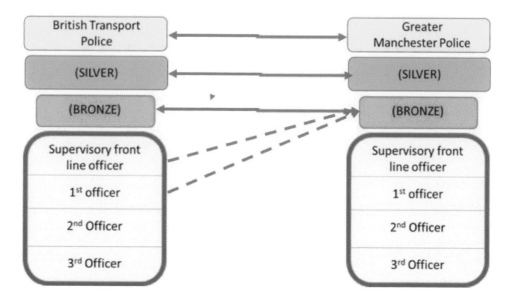

Even with this simple example, it is apparent from a brief look at the diagram that if the personnel (connected by dashed lines) upon whom duties fell had:

- a personal card indicating who should be contacted
- listed protocols dictating what information should be conveyed when contacting
- the results of a prior analysis showing what spare communication device should be provided
- DGRs stating the priority order of communications both to and from the individual that should occur
- A LEDSM diagram of their Enterprise and the roles in other Enterprises that they must contact.

then the METHANE message required would almost certainly have been sent. All that was required was two officers present with different communication devices, (e.g mobile phone and radio), which a well conducted LEDSM study would have plainly shown were required.

The 'granularity', 'getting down in the weeds', or 'the devil is in the detail', whatever expression one wishes to use as a description, is where most failures will occur. Therefore that is the area that needs to be approached more cautiously using the techniques here than the current basic overconfident assumption that the command levels will have perfectly organised their Enterprises.

The diagram above shows a LEDSM interpretation of communication the official report describes. However, after a LEDSM cycle show in the first figure had been completed a much better picture emerges. It could have been agreed at the Resilience Forum that any officer may send a METHANE message if they know their immediate superior front-line officer, who would normally send the message, is incapacitated. This is little different from the questioning by the ambulance service of a random member of the public calling about the emergency so it is perfectly acceptable to delegate the task. The next diagram shows the hypothetical situation of the front-line supervisory officer being incapacitated and the METHANE message being sent by the 1st Officer. Everyone involved at the front needs to know so the flow of data from the 1st Officer may be something similar to that shown.

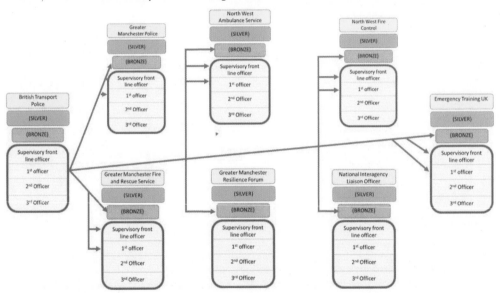

Establishing the Necessary Communication Systems

The systems that will be required to communicate may need a deep analysis as certain environmental conditions that may occur and the uniqueness of internal communication processes and systems could make it impossible to communicate with other organisations.

Any system used in the internal network or that is intended to be used in communication with external organisations needs to be reviewed. Typical environmental conditions affecting communication may be the difficulty of getting signals from other systems in the network, such as when in high rise towers, as well as problems due to mobile phone networks having trouble with the volume of traffic that can grow exponentially in an emergency situation, dedicated lines being a necessary fallback where common public communication systems fail.

Generally, for emergency communications, a minimum of two paths should be available for information to flow within an Enterprise or across to other Enterprises. Each Enterprise needs to be clear about what systems they intend to use for internal and external Enterprise relationships and have confidence that they are compatible with other Enterprises.

Development of the LEDSM diagram to include other Stakeholders

The next stage of the development of the Protocols and DGRs is to include in the LEDSM diagram the other Enterprises. In a computerised system, (yet to be developed), the LEDSM diagrams, with associated protocols and DGRs would all be available to any depth or granularity as required from a single screen by zooming in and zooming out.

Print-outs so individuals have with them a LEDSM picture of their connections, protocols to adhere to and DGRs that are expected to be fulfilled would also be possible but it is simple to do that on paper for individuals, the complete computerised picture will be much more complex but, importantly, verifiable.

The Second and Subsequent Meetings to agree the LEDSM model, Protocols and DGRs

Before each subsequent meeting of, in this case The Resilience Forum, each Enterprise will go through the LEDSM cycle. All improvements and changes required to communication protocols should be taken to the Resilience Forum. Requests alone should be considered insufficient as each change should have a rationale that, it is patently obvious to the other Enterprises, is an improvement to the overall intention of saving lives. Suggested DGRs should be negotiated because one Enterprise's idea of how quickly they need information may be unrealistic to the supplying Enterprise. Ultimately the best possible functioning of the emergency services will emerge.

Declaration of the Refined Communication Protocols and Dependency Guarantee Relationships with Necessary Training

When the LEDSM has been developed through sufficient iterative use of the process cycle, both within an Enterprise and at, in this case, the resilience forum, there will be greater confidence among the Enterprises and especially those personnel working for the Enterprise at the front line levels, that communication associated with any emergency situation will be adequate and certainly at a much lower risk of failure at any point in the rescue mission.

The output should include declaration of

- appropriate communication devices in both type and numbers
- protocols acting as instructions to cope with all anticipated problems
- protocols derived generically for unanticipated problems for each level in the Enterprise
- DGRs on the critical aspects of the Protocols concerning information that should be supplied within certain time frames
- LEDSM diagrams, at both the macro level for the management of Enterprises, and micro level, to graphically illustrate the communication links the front line personnel should have.

A Protocol for the BTP Officer

- Call for back-up assistance from other BTP officers and request First Aid equipment be brought including items from for pain relief from secured cabinets.
- Within 2 minutes of arrival at scene that can be described as a <Major Incident>, declare the Major Incident text message to all Enterprises in the Emergency Response WhatsApp group using your Major Incident mobile phone
- Within 4 minutes of arrival at a scene that can be described as a <Major Incident> send a METHANE text message to all Enterprises in the Emergency Response WhatsApp group using your Major Incident mobile phone.
- If no Enterprise acknowledges either message within 1 minute of sending, use your standard radio to establish contact with either the BTP Bronze Level Officer or a Bronze Level Officer in another Enterprise and ask them to send the two messages and request confirmation that acknowledgements have been received.
- In the event of all emergency communication systems failing, send a colleague to report the event by any means available including personal attendance at a coordinating response centre or use of public phones.
- When communications established and METHANE and <Major Incident> messages sent try to assist injured and wait for the declaration of Rendezvous Point, (RVP), by the NWAS, in order to direct anybody helping with evacuation of casualties to the NWAS RVP.
- If no RVP received from NWAS within 4 minutes of successful METHANE and <Major Incident> communication, request another Enterprise declares the best RVP for all parties using the whatever communications are available but preferably the Emergency Response WhatsApp group.

Above is what a Protocol may look like, though this is drawn up for illustrative purposes and is only part of the duties expected.

Conclusion

It is suggested that, just as safety-critical software requires a degree of formality in requirements capture and design, data communication requirements prepared for emergency situations should be developed more formally than is currently practiced. If LEDSM and associated techniques are used by a Resilience Forum, or indeed by any group of organisations planning for emergencies and safety-critical control, the overall response may be better, and may improve the outcomes for some of the victims.

Image Attribution
front picture: Tomasz Kozlowski licensed under creative commons BY-SA v4.0

References

[1] The Layered Enterprise Data Safety Model (LEDSM) - A Framework for Assuring Safety-critical Communications – Nicholas Hales - [SCSC-176] Safety-Critical Systems eJournal vol.1 no.2 Summer Issue 2022

[2] Manchester Arena Inquiry -Volume 2: Emergency Response Volume 2-I Report of the Public Inquiry into the Attack on Manchester Arena on 22nd May 2017 - Chairman: The Hon Sir John Saunders - November 2022

Nicholas Hales

Nicholas is a retired engineer who as a student worked in communications with Royal Naval Reserve. He has a degree in Computer and Control Systems and Masters degrees in Modern Electronics and Quality Improvement and Systems Engineering. He has been involved in software safety and numerous projects involving safety-critical avionics for the MoD. He is a member of the IET.

Avoiding Safety and Cybersecurity Risks in Autonomous Systems

ISO 26262
Road Vehicles - Functional Safety

ISO 21434
Road Vehicles - Cybersecurity Engineering

Dhanabal Arunachalam discusses a proposed safety analysis technique that allows synergy between safety and cybersecurity during the development of partial, conditional, high and full driving automation vehicles. He provides an overview of the technique and the analysis concepts where safety measures are identified and cybersecurity conflicts are resolved. He also identifies and addresses the challenges confronted when conducting the analysis during development of hardware, software and the overall system.

Safety and cybersecurity engineering are fundamental areas upon which autonomous vehicle design should be focused on risks (both safety and cybersecurity) related to driving. Both disciplines are widely used at various stages of vehicle development – to assist in design, verification, and interfacing with other software of vehicle electrical/electronic systems.

Safety and cybersecurity analysis should go together during the product development lifecycle. For example, in a connected car, a safety analyst provides proper safety measures to deploy an airbag during a crash. However, the cybersecurity analyst provides cybersecurity measures for denial of airbag deployment if an external attacker has taken controllability of the airbag system, thus leading to potential conflicts. Hence, to address this problem, it is necessary to have synergy between the disciplines, especially during safety and cybersecurity risk analysis.

The following image illustrates that when safety and cybersecurity analysis are executed independently, we see that the different discipline mindsets don't help in reaching the overall goals of achieving a sufficiently safe and secure autonomous vehicle.

Different mindsets can emerge as safety is concerned with inadvertent unintended behaviour, whereas security deals with goal-seeking opponents.

To ensure safety and cybersecurity are synergised as shown in the figure, the following proposed technique allows safety and cybersecurity team to work together to achieve autonomous safety and cybersecurity intended objectives.

Safety and Cybersecurity – scope

The goal of this article is therefore to provide a concept overview and process workflow for a proposed safety analysis technique for developing harmonised safety and cybersecurity vehicle functions.

The starting point for this is the extraction of concepts from the automotive safety standard "ISO 26262" [1] and cybersecurity standard "ISO 21434" [2]. These standards are the foundation of this article, as they reflect current best practice in managing safety and cybersecurity risks for the automotive industry.

Safety and cybersecurity measures should go hand-in-hand to achieve state-of-the-art solutions in an autonomous vehicle.

Safety and Cybersecurity – ISO standard overview

Relevant extracts from ISO 26262 about cybersecurity in the concept phase are as follows:

- cybersecurity threats to be analyzed as a hazard from a functional safety perspective in order to support the completeness of the hazard analyses and risk assessment and the safety goals
- functional safety can provide information such as hazards and associated risks to support the cybersecurity identification of threats

Note that ISO 26262 does not address cybersecurity threats, however, it provides guidance on potential interaction of functional safety with cybersecurity.

Relevant extracts from ISO 21434 about functional safety cover the interdisciplinary exchange of the following activities:

- threat scenarios and hazard (ISO 26262-1:2018 [1]) information
- cybersecurity goals and safety goals (ISO 26262-1:2018 [1])
- cybersecurity requirements conflicting or competing with functional safety requirements (ISO 26262-1:2018 [1])

Functional safety should therefore consider cybersecurity threats with safety impacts, identified through security analysis, and make corresponding additions to the safety analysis. Cybersecurity should look at the safety related hazards identified through safety analyses and if it is possible for an external attacker, to trigger the safety faults in the system through external manipulation.

Safety and cybersecurity engineering are fundamental areas upon which autonomous vehicle design should be focused on risks related to driving.

Risk analysis process overview

Safety and cybersecurity functions play a crucial role in determining hazards that occur due to violation of safety and/or cybersecurity goals, as shown by the processes in the top section of the figure.

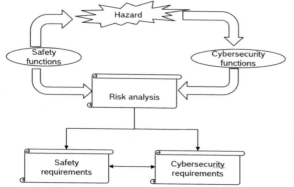

These functions are then further assessed (lower part of the figure) through risk analysis management, which defines safety and cybersecurity requirements, and it is evaluated via risk analysis (ie. The proposed safety analysis technique), to identify critical safety, cybersecurity, and liability risks.

For example, in connected car camera/radar systems, an attacker can either activate automatic emergency braking (AEB) when not needed, or make the brakes fail when needed. The proposed analysis will help derive safety & cybersecurity requirements on intended activation of AEB.

Proposed safety analysis - workflow

The proposed safety analysis performs safety and cybersecurity risk management in a harmonized manner as depicted in the following diagram.

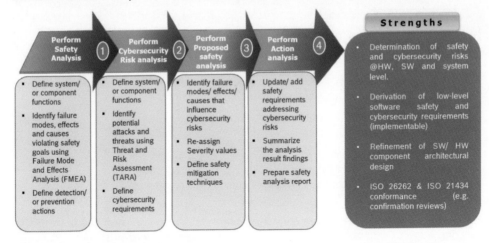

Steps 1 and 2 of the diagram defines safety and cybersecurity risk management processes that safety and cybersecurity teams execute independently. From Step 3 onwards, core attributes associated with cybersecurity failures are mutually aligned with the safety team and safety mitigation strategies are defined to address those cybersecurity risks. As part of the safety mitigation strategy, measures like degradation states, fail safe and fail operational concepts are refined based on the cybsersecurity risk mitigation strategy.

Detailed workflow diagram of proposed safety analysis

The detailed activities of Step 3 from the above proposed process are illustrated in the following diagram.

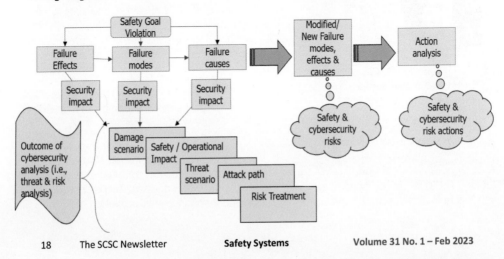

This process starts by considering safety risks from potential attacks and vulnerabilities, which are obtained via damage scenario and threat scenarios.

Initially, failure modes impacting safety goals are determined via typical Failure Mode & Effects Analysis (FMEA). On the security side, damage scenarios and threat scenarios are identified via Threat Analysis and Risk Assessment (TARA). Next, Failure modes are checked against each of the damage scenarios, threats, and attack paths. As part of this check, critical points like violation of safety goal by damage scenarios, threats, and attack paths are taken into consideration. This verification activity is repeated for each of the failure effects and failure causes. As an outcome, it is possible that new hazards and safety goals will be identified. As a post verification activity, action analysis is executed upon which all occurrence and prevention measures are re-analysed. Also, the severity level of safety effects is revised, and risk priority numbers are modified accordingly. As an outcome of this analysis, the safety mitigation strategy is refined in a way that cybersecurity risk strategies are integrated.

Proposed safety analysis - benefits

The proposed safety analysis offers a wide variety of merits e.g., well-trusted and effective processes, customized use of standardised safety analysis tools and methods to ensure end-to-end adherence with functional safety and cybersecurity standards, compliance against product safety/ or cybersecurity process with an established functional safety and cybersecurity intent, assumptions/ limitation during execution of proposed safety analysis, conform to cohesive safety/ cybersecurity culture and traceable documentation, etc.

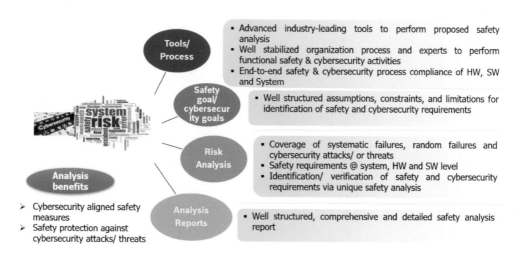

Proposed safety analysis - summary

The proposed safety analysis technique allows safety and cybersecurity problems to be addressed during various stages of the safety development lifecycle and it provides a state-of-the-art solution for the major issues faced during autonomous vehicle development. A unique failure mode analysis along with guidelines and processes are followed during execution of the proposed safety analysis for system, hardware, and software development. This ensures that both safety and cybersecurity issues along with vehicle scenarios are analysed and corresponding cybersecurity aligned safety measures are identified.

References

[1] ISO 26262: 2018   Road vehicles – Functional safety, Part [1], [3], [4], [9]

[2] ISO 21434   Road vehicles – Cybersecurity engineering, Chapter [8], [9], [10]

Dhanabal Arunachalam, Functional Safety Expert, Robert Bosch LLC, USA

Dhanabal has spent over 16+ years working on functional safety engineering for automotive and railway systems. His functional safety expertise includes hazard analysis and risk assessment, FMEA/FMEDA, FTA, DFA, tool qualification, software component certification support, SOTIF analysis, safety case & safety manual preparation and safety-cybersecurity interaction analysis. He has a rich experience in harmonized process & guideline derivation for functional safety and cyber security activities and has been involved in a number of safety related projects for achieving compliance of EN50126, EN50128, ISO 26262, ISO 21448 & ISO 21434 standards.

Data Safety Analysis using RADISH

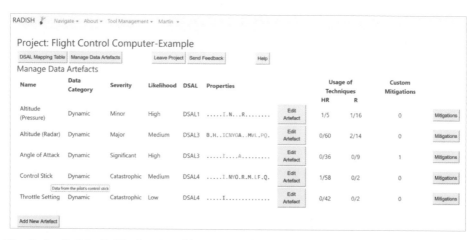

The Data Safety Initiative Working Group (DSIWG) is now in its 10th year of operation and its Data Safety Guidance document has been helping engineers assess and manage data safety risks. Divya and Martin Atkins discuss a new tool that they are developing to help practitioners implement the guidance by automating many of the processes in the guidance itself.

The Data Safety Initiative Working Group (DSIWG) was the first working group to be set up under the Safety Critical Systems Club (SCSC), in 2013. With 88 members and 74 meetings later, it is still going strong, as data becomes increasingly important in safety-critical systems with every passing year. Over this period, the DSIWG has published and refined the Data Safety Guidance (Guidance) – now at version 3.5 [1], into a mature and well-regarded document, which is increasingly used in industry, and referred to by industry-specific guidance, such as the UK National Health Service's clinical risk management standards: DCB0129 [2] and DCB0160 [3], and some international safety standards, such as the upcoming version of IEC 61508 [4].

RADISH (Risk Assessor for Data Integrity and Safety Hazards) is a software tool being developed by Mission Critical Applications Limited (mca-ltd.com), to assist a data safety practitioner developing a data safety case using the Guidance, by:

- recording the decisions that are made
- automating parts of the data safety assessment process
- helping the practitioner to choose between the risk mitigations that are recommended by the guidance, given the nature of each risk

This article provides an overview – for more information, and access to the RADISH tool, visit data-safety.tech/tooling

Data Safety

Modern Systems use data to make safety-critical decisions. Errors in, or the incorrect use of such data, can cause harm to life and the environment. Ensuring the safe use of data is a complex challenge faced by all industries, but some industries, such as healthcare, are particularly reliant on data. The risks from data will only increase as our systems become more inter-connected, autonomous, and driven by data-intensive technologies such as the Internet of Things, Artificial Intelligence and Machine Learning.

Accidents are happening...

2019 - Immensa Labs False Negative Covid-19 PCR Tests

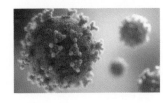

- An estimated 1,000 deaths.

False-negative Covid test results meant that 43,000 people were not told to quarantine, further propagating the virus.

2018/19 – Boeing 737 MAX

- Lion Air Flight 610, 189 lives lost
- Ethiopian Airlines Flight 302, 157 lives lost

No redundancy of critical angle-of-attack data to the MCAS system. Also inadequate training materials, missing in-service problem reporting, and inadequate responses to failed test reports.

2017 – Irish Search and Rescue Helicopter

- Lost with all crew

Flying in zero visibility, the helicopter flew into a hill that was not in the map data loaded into its Enhanced Ground Proximity Warning System.

The Data Safety Guidance

The DSIWG publishes cross-sector best practice in the Data Safety Guidance. The Guidance describes a Data Safety Management Process, which can be integrated into an overall Safety Management System. A major part of the process involves considering appropriate mitigation techniques for each safety-related data artefact, based on the criticality and other metadata characterizing each data artefact.

RADISH: Risk Assessor for Data Integrity and Safety Hazards

Mission Critical Applications have been members of the DSIWG since 2017. They realised that the highly table-driven process to choose mitigation techniques for risks was difficult (and error-prone) to apply, and very suitable for automation. This was the basis of the development of the RADISH tool to guide a data safety practitioner through this part of the data safety process.

The formalisation of the Data Safety process needed to implement RADISH is also feeding back clarifications to the Guidance, and formalising aspects of the process.

Grant funding from the Lloyd's Register Foundation made it possible to produce a proof-of-concept tool, and that is now being progressed by funding from Innovate UK.

The RADISH Tool in the context of a large development project

The RADISH tool is a central repository of information about the Data Safety case for a development project. Data Safety engineers following the Guidance, identify the data artefacts in the system, and the safety properties that are important for each artefact. The engineer chooses which mitigation techniques to use out of those recommended, or highly recommended by the guidance, adding those to the requirements of the system.

During the design and development process, RADISH can generate reports showing the risks that have mitigations, and those where more work is needed, giving project management a view of the state of the data safety process.

When the analysis is complete, RADISH can generate a report that can be included in a Data Safety Case to support the safety argument.

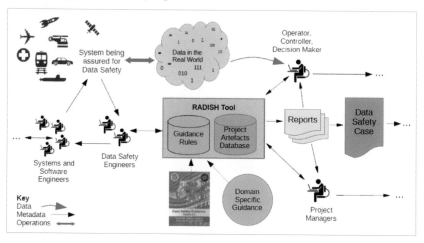

RADISH in Practice

RADISH is a web-based application supporting the collection, management and maintenance of information about the safety of the data assets of a project. It records the identified data safety risks from each data artefact.

The tool also *suggests* mitigation techniques available from the Guidance, to improve the trustworthiness of the data. All risk mitigation decisions are captured along with supporting *justifications*, for inclusion in the Data Safety Case.

These steps can be seen in the screenshots below:

The first shows the creation of a new data safety artefact along with the properties of the data that need to be maintained for safety. The second shows an example of a recommended mitigation arising from the Guidance and automatically displayed by the tool.

References

[1] Data Safety Guidance, SCSC DSIWG, Version 3.5, available to download at scsc.uk/scsc-127H
[2] DCB0129: Clinical Risk Management: its Application in the Manufacture of Health IT Systems
[3] DCB0160: Clinical Risk Management: its Application in the Deployment and Use of Health IT Systems
[4] IEC 61508 Functional safety of electrical/electronic/programmable electronic safety-related systems

Image attributions
Covid © Fusion Medical Animation | unsplash.com
Lion Air Flight 610 © PK-REN, CC BY-SA 2.0 | wikipedia.org,
Ethiopian Airlines Flight 302 © LLBG Spotter, CC BY-SA 2.0 | wikipedia.org
Irish SAR Helicopter © Riatsnapper, CC BY-SA 3.0 | wikipedia.org

Dr Divya Atkins, Managing Director, Mission Critical Applications Limited

Divya Atkins (née Prasad) has a background in Computer Science, and has conducted research in formal methods, high-integrity and real-time systems and software metrics. Her experience includes safety-critical development, and project management, and she is interested in computer and network security.

Dr Martin Atkins, Technical Director, Mission Critical Applications Limited

Martin has a research background in Object-Oriented languages, and software development experience in operating systems, embedded, safety-critical systems, and software tools. He has also managed large networks, developed hardware, taught courses and undertaken technical authoring.

Safety Futures Initiative
Introducing ...
Laure Buysse

The **Safety Futures Initiative (SFI) has been established to provide support and guidance for those in the earlier stages of their career in systems safety. The initiative aims to help develop the next generation of safety engineers so that club members can continue to make systems safer well into the future.**

The SFI has now had several events such as "Get To Know You" meetings, podcasts and there are plans for a lecture competition this year. But what is it actually like being a member of the SFI? What has the SFI journey been like and how is SFI making a difference?

Laure Buysse, a member of the SFI, kindly agreed to introduce herself and provide her thoughts on her career, the SFI and the future!

I began my journey into system safety when I started my studies in engineering technology at KU Leuven. I quickly discovered my interest was much broader than just software development and longed to know more. This urge to learn, led me to simultaneously complete a Master of Engineering Technology in ICT engineering and a Master of Engineering Technology in Electronics Engineering.

My passion for research resulted in two internships centred around deep learning. It is no surprise that during this time, the topics of "Safe AI" and "System Dependability" kept surfacing as major issues and research pillars. Especially within the domain of autonomous systems, such as self-driving cars and unmanned aerial vehicles, where safety remains a heavily discussed topic. My interest was piqued and I started to delve deeper into the world of safety.

In September 2021, I became a researcher at KU Leuven under the guidance of Prof. Pissoort on the TETRA Project Safety Assurance 4.0. This gave me the chance to fully focus on safety and delve deeper into topics such as safety cases, safety standards, Systems Theoretic Process Analysis (STPA) and more. I have since guided several projects around these topics and was able to disseminate some of the knowledge among the local industry and some of my peers. The picture shows me presenting my work during the International Electronics-ET2021 conference in Sozopol, Bulgaria (September 2021).

More recently, I travelled to the UK for five weeks as a visiting researcher of the Assuring Autonomy International Programme (AAIP) at the University of York. Currently, I am pursuing my PhD at KU Leuven as a FWO fellow thus continuing my research on the safety of autonomous systems with a focus on (executable) safety cases and digital twins.

I hope for the SFI to be part of this journey to learn and grow and build a community of young like-minded individuals.

Although my journey has been limited so far (with just a get-to-know-you event), my hopes for the initiative are big. This past year it has become abundantly clear to me that safety is a team sport, where the right information, discussion and community is vital.

Moreover, while young technologist such as me are highly trained in technical engineering topics, information about general system safety is rather limited. I hope for SFI to provide the space and opportunity to discuss all things safety and share ideas and experiences across industries, creating a real community of inspired and enthusiastic people.

> "It has become abundantly clear to me that safety is a team sport, where the right information, discussion and community is vital".

As part of the bigger SCSC, the SFI is probably the perfect starting point to bring young professionals from both academia and industry together. While many of us may have a different background, ranging from pure safety to software engineering and more, I don't believe this to be disadvantage. On the contrary, though initially it might be difficult, there is undoubtably immense value in bringing all these backgrounds and expertise together.

Laure Buysse is a FWO fellow at KU Leuven studying for a doctorate researching the safety of autonomous systems with a focus on (executable) safety cases and digital twins.

Membership

Membership for the SFI is free for the first year, so please sign-up if you would like to get involved (please see www.scsc.uk/membership).

SFI members get access to all SFI events and activities, as well as discounted fees at SCSC Events.

Further Information

If you would like further information about any aspect of the SFI, please do get in touch with Zoe Garstang (zoe.garstang@scsc.uk) or Nikita Johnson (nikita.johnson@scsc.uk).

The Future of Requirements for Safety Critical Systems

The "Future of Requirements for Safety-Critical Systems" seminar was held on the 22nd September 2022 at the Chartered Institute for IT, London and virtually online. The seminar looked at the importance of requirements and specifications for safety-critical systems, the state of the art in requirements management and what can go wrong if requirements are not correct in some way, e.g. missing, ambiguous or incorrect. Bernard Twomey provides a summary of the event.

Mike Parsons introduced the seminar and seminar speakers and pointed out that a large number of failures could be traced back to the requirements, which in some cases were non-existent.

Requirements – How Not to get Them Badly Wrong...

Jane Fenn, BAE Systems, provided an insightful journey into the world of requirements and how defects at the requirements phase could have a significant impact on the demonstration of safety to the relevant regulatory authorities.

Jane stated that "it is never too early to start thinking about safety and isn't something you do at the end of the program" and then explained the principles of 'Architectural Trade Off', Functional and non-Functional safety, Validation of Requirements, and the use of Tool Supported Validation of Requirements.

Key points from the presentation included:

automated testing could result in requirement errors not being identified'
- There is no perfect solution
- do not be overly seduced by tools
- significance of getting the requirements right is going up...significantly.

Aircraft Accidents that have resulted from Incorrect Requirements

Dewi Daniels, Software Safety Ltd, provided a very structured presentation which started with a positive statement that:

'Not a single hull-loss accident in passenger service has been ascribed to failure of software to meet its requirements'. His next slide however, stated that *'There have been several accidents where the software behaved exactly as specified in its requirements, but those requirements specified unsafe behaviour in some unforeseen circumstances.*

Dewi presented several accidents, many of which resulted in fatalities. His presentation clearly described the accident causes with one case study raising the question on *'who is responsible for providing accurate information to the pilot when landing?'*

He stated that the number of lines of code is increasing, and the use of outsourcing to software houses is a potential risk, as they may not have sufficient design experience. He also raised the issue of a loss of independence in the verification process.

Key points from the presentation included a realisation that the avionics industry is (surprisingly) good at deploying correct software, accidents have been caused by requirements errors e.g. the Maneuvering Characteristics Augmentation System (MCAS) software attributed to the Boeing 737 MAX accidents of 2018/2019 (see Boeing 737 MAX Safe to Fly? – Safety Systems Newsletter, Volume 29, Number 1) behaved exactly as specified in its requirements, and in one case the accident report blamed the pilots.

The presentation highlighted the necessity to get better at writing and validating requirements if we are to improve aircraft safety and prevent loss of lives.

Tackling the Uncertainties in Requirements

Nick Tudor, DRisQ Ltd, provided an insight into the development of automatic Formal Methods based verification technologies. His presentation identified the significant costs associated with verification activities which then led onto the need for requirements attributes to be 'Clear' and 'Precise', which were supported with a number of excellent examples.

Nick then went on to explain the problem in communicating our requirements to specialists and non-specialists alike. He stated that writing clear and precise requirements should be taught at university, as everything goes wrong from poor requirements capture. This was again, supported with excellent examples along with potential solutions, such as, adding 'Universal Quantification' that would help in defining a reasonably unambiguous statement of requirements.

Nick then explained the 'Tools' developed by DRisQ and how they could be used in the verification of requirements. Kapture – designed to automate compliance to a requirement standard. CAESAR provides a formal link between system requirements and software requirements, Modelworks verifies the design against Kapture, ClawZ

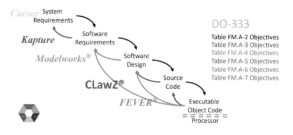

automatically verifies the auto-code and FEVER provides automatic verification of the object code.

Key points from the presentation included a statement that '*nothing can tell you that you have the right requirements and there still needs to be a human review*', '*no tool can do everything for you*', and developing requirements is a social exercise and you should not expect them to be right first time.

One Million Monkeys or one Bored Teenager: The Challenge of Security Requirements

James Yolland, Ebeni Ltd, discussed the world of security and the development of security requirements. His presentation started with examples of security failures and how the threats are real and potentially harmful e.g. Vishing and virtual kidnappings. He then went onto explain the foundations for good security requirements relies on identifying your attackers capabilities and their motivations along with excellent knowledge of the system and the stakeholders through life.

James explained the goal of security is to protect the system and the stakeholders from unacceptable harm. He explained that following a successful general threat/hazard attack the risk largely remains unchanged with a clear recommendation that security requirements would need to be updated.

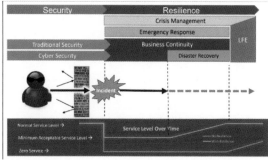

The presentation then proposed a recommendation for writing better security requirements, which included, aligning and engaging with the systems engineering lifecycle, understanding the systems and the stakeholders, identifying threats, vectors and critical assets and addressing the full lifecycle of the system.

Key points from the presentation included a realisation of the complexity of the systems domain and the necessity to have accurate knowledge and identification of the critical assets along with early engagement with the systems engineering lifecycle, preferably at the start of a project, not at the end.

Requirements Management for Safety Critical Systems

Steve Rooks, IBM, explained that poor requirements management has a significant impact on your business in the way of requirements rework, project delays and project impacts, citing a wide range of statistics to support his argument.

He explained that standards initially increase project costs due to 'inadequate level of detail and process for requirements, improper tool qualification (too much or too little), and weak process and checklist management.

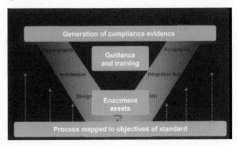

Steve then explained the principles of automating safety-critical compliance, linking of data to support compliance, requirements management fundamentals for safety-critical systems development and data models that include safety and cyber security.

Steve then explained the principles of IBM DOORS Next, FMEA and presented an excellent demonstration on how the tool has been used (e.g. Infusion Pump Requirements) and covered link validity, traceability and the use of an open link data set – Open Services for Lifecycle Collaboration (OSLC).

Configuration management was discussed which covered audit trails, variants, global configuration management, requirements validation and reporting.

Key points from the presentation included an acknowledgement that poor requirements capture is a risk that can be reduced by the use of IBM DOORS and management of requirements can be of benefit for many safety-critical projects along with helping to improve quality and productivity.

Requirements Management

Sean Sullivan from 3DS Dassault Systemes, gave his presentation online and explained the principles of requirements engineering and the need to consider hardware and software together. He stated that the relationship between Elicitation (vision and scope, identified use cases), Analysis (data dictionary), Specification (derived requirements) and Verification (requirement verification) are extremely important.

The necessity for requirements to be under configuration management control during all steps of the lifecycle offers extremely valuable services to the end users. The message from Sean aligned with all the other presentations on the need for requirements capture and validation to be taken seriously by all relevant stakeholders.

The day finished with a lively panel discussion with a common theme of requirements that are clear and precise can help prevent accidents and could improve costs.

It was felt the day went very well with interesting and very useful presentations, whether the attendees will board an aircraft again is debatable!

RAF Museum Midlands (Cosford) Tech Trip Report

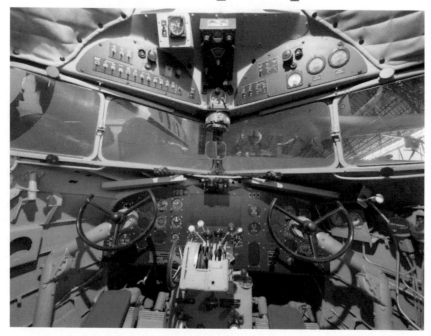

After the success of the inaugural SCSC Tech Trip to Bletchley Park, the SCSC's next organised trip was to the RAF Museum Midlands (Cosford) on Saturday 19th Nov 2022. The trip, which was open to everyone included a 3-hour guided tour of the museum's collection. Dave Banham provides a report of the trip.

Our guide was Cameron, a former engineer from Bentley cars and who had nurtured a lifelong interest in aircraft as his father had been a test pilot, and now retired, is learning finally to fly a small aircraft! He was certainly very knowledgeable on the collection and talked us through it for 3 hours non-stop.

Cosford is relatively close to me and I have visited it numerous times with family and growing children, but never with a guided tour. The difference a knowledgeable guide brings to understanding and contextualising the technology of each exhibit in its era and with that era's motivations is considerable!

Our tour covered the Test Flight, War in the Air (WW1 and WW2), and the Cold War museum buildings. On learning that his tour party were safety engineers Cameron pointed out a few safety-related points of interest, such as the notice next to the Mk1 Spitfire's engine hand-crank socket that reads:

"HAND TURNING GEAR FOR MAINTENANCE ONLY. IF USED FOR EMERGENCY STARTING AIRCRAFTSMAN MUST HAVE ROPE FROM HIS WAIST TO THE UNDERCARRIAGE TO PREVENT HIM FALLING INTO THE AIRSCREW."

I was also bemused by the Messerschmitt Me 262A-2a Schwalbe's jet engine starter motor – it was a single cylinder "lawnmower" engine with a cable pull start! But back in the day that must have been an innovation that freed the jet engine from needing ground starting equipment. Just imagine reaching in and pulling the starting cable for a jet engine!

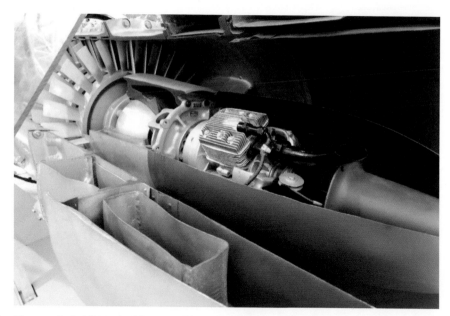

The Museum's Cold War building is relatively modern having been purpose designed for the collection and opened in 2007. There's a staggering array of aircraft, ground vehicles (including some tanks) a helicopter (a Sikorsky transport), and several missiles including the huge Douglas Thor Intermediate Range Ballistic Missile (IRBM) – some 20m tall.

The high-altitude bombers were the Valiant and the Victor, but they were soon surpassed by lower altitude fast bombers such as the Vulcan, but remained in service by being repurposed as refuelling tankers. The Victor had three mid-air refuelling pods; two for fast jets and one for a large-bodied aircraft.

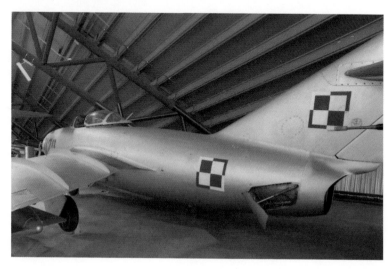

The slender English Electric Lightning has pride of place showing off its polished metal work on all sides with its vertical display as the picture on the right shows. (And alongside it, just visible bottom left, is a Gloster Javelin and on the right, behind the Lightning is the Douglas Thor).

All in all it was a good, but exhausting, visit and one that is highly recommended. The museum is free to enter, although donations are encouraged and there is a fee for a guided tour and for the car park. There's a railway station close by too.

The museum's website is: https://www.rafmuseum.org.uk/midlands/ and it has extensive information about the exhibits too.

Photographs copyright © Dave Banham and Dewi Daniels.

Seminar: Safety of Autonomy in Complex Environments

THE SAFETY-CRITICAL SYSTEMS CLUB, Seminar:

Safety of Autonomy in Complex Environments

Thursday 20 April, 2023 - London, UK and blended online

This 1-day seminar will consider the safe use of autonomy in complex environments (for example a self-driving vehicle in a city environment), featuring (i) work undertaken over the last couple of years at the AAIP at the University of York. This work has produced a framework document, "Guidance on the Safety Assurance of Autonomous Systems in Complex Environments (SACE)" outlining techniques and approaches for assurance and gives an example safety argument, (ii) outputs from the SCSC Safety of Autonomous Systems WG and (iii) work undertaken by Roger Rivett in the automotive domain.

Speakers include:

TBC, SCSC SASWG

Richard Hawkins, AAIP

Roger Rivett, Consultant and AAIP

Further details TBA.

This event will be held at the TBC, London and also online via Zoom.

Seminar: How to Write Compelling Safety Arguments

THE SAFETY-CRITICAL SYSTEMS CLUB, Seminar:

How to Write Compelling Safety Arguments

Thursday 8 June, 2023 - London

This event will consider what makes a compelling safety argument, and how to use narrative techniques to improve the flow and 'story' of the argument. Talks include one from a professional writer who will explain how to write prose so as to engage and grip the reader.

There will be discussion on the use of metaphor and imagery in arguments and a workshop in the afternoon.

Further details TBA.

Speakers include:

TBA, Ebeni

Catherine Menon, University of Hertfordshire

60 Seconds with ...
Prof. Philip Koopman

Phil is an internationally recognised expert on Autonomous Vehicle (AV) safety whose work in that area spans over 25 years. He is also actively involved with AV policy and standards as well as more general embedded system design and software quality.

His pioneering research work includes software robustness testing and run-time monitoring of autonomous systems to identify how they break and how to fix them. He has extensive experience in software safety and software quality across numerous transportation, industrial, and defence application domains including conventional automotive software and hardware systems. He was the principal technical contributor to the UL 4600 standard for autonomous system safety issued in 2020.

He is a faculty member of the Carnegie Mellon University Electrical and Computer Engineering department, where he teaches software skills for mission-critical systems. In 2018 he was awarded the highly selective IEEE-SSIT Carl Barus Award for outstanding service in the public interest for his work and in 2022, he was named to the National Safety Council's Mobility Safety Advisory Group. He is the author of the book *How Safe is Safe Enough: Measuring and Predicting Autonomous Vehicle Safety* (2022).

What first attracted you to working in the field of System Safety?

It has been more collecting pieces over time than a conscious plan to work in the field of System Safety. As a teenager I taught bicycle safety to school kids as a public service project. After college, wearing a personal radiation dosimeter 24 hours a day in a sealed environment during my Navy submarine service certainly focused my attention on safety.

I got an opportunity to be the dependability/safety expert on an early autonomous vehicle project at Carnegie Mellon University in the mid-1990s by being the right person at the right place at the right time. Since then I've been fortunate to have opportunities to work with safety in various domains including rail, industrial controls, commercial buildings, consumer electronics, power systems, and conventional automotive by doing industry design reviews. I have also done some product defect litigation expert work.

What aspect of your career are you most proud of?

For me it has always been about finding a way to make a difference. Most recently, I think leading the development of the ANSI/UL 4600 autonomous vehicle safety standard has pushed the conversation forward on autonomous vehicle standards and safety cases in both the technical and public policy arenas.

The professional recognition I've received that is most important to me is the IEEE SSIT Carl Barus award for my work on computer-based system defects in automotive applications. Receiving that award is inherently bittersweet because it is only given to people who persevere amidst controversy, typically involving public safety.

What advice would you give to yourself age 12?

If it is worth doing, it is worth taking your best shot.

What excites (or frightens!) you the most about the future of Autonomous Vehicles?

The deployment governance model, especially in the US, is dominated by big money pressing for quick results. This pressure has done short-term harm via needlessly increasing the risk of individual loss events. But just as importantly, long-term harm is being done in the areas of undermining regulatory authority, exclusion of legitimate stakeholders from deployment decisions, and eventual disillusionment with the technology when hype collapses.

What's your most favourite quote or motto?

Illegitimi non carborundum.

If you could learn to do anything, what would it be?

The next thing I need to learn to continue on my path. Every day is a good one if I'm learning.

What's the best piece of advice you've ever been given?

Find a way to share your talents.

In a world of conflict, austerity and climate change, which work of art or fiction best sums up your current outlook in life?

I posed that question as a prompt to "DALL·E 2" (a new AI system that can create realistic images and art from a description in natural language), and it generated this "work of art" on the first try.

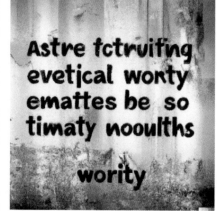

Here is an "AI" (really it is not intelligent at all) that generated a complete nonsense response to a reasonable question. But the nonsense is captured in a format that people are accustomed to accepting as deep wisdom, and might even pass if you don't look long enough to try to figure out what language it is written in. (Google translate says this is not a language it knows. But that is an "AI" too I suppose).

I think there is a cautionary message here for those who argue that machine learning will inevitably be a superior replacement for human judgment in any particular application.

Recent Publications

Guidance on the Safety Assurance of autonomous systems in Complex Environments.
http://www.york.ac.uk/assuring-autonomy/guidance/sace/

Revealing a newly discovered fundamental law of inequality.
http://www.amazon.co.uk/Exposing-Natures-Bias-Clockwork-Universe/dp/1908422041

Safety-Critical Systems eJournal vol.1 no.2
Summer Issue 2022
scsc.uk/scsc-176

Proceedings of the 30th Safety-Critical Systems Symposium.
scsc.uk/scsc-170

Guidance on the management of data safety risks.
scsc.uk/scsc-127H

Three decades of work in the field of safety-critical systems as told through the SCSC Newsletter.
scsc.uk/scsc-169

Guidance on assuring autonomous systems.
scsc.uk/scsc-153B

Guidance on assuring safety-related services.
scsc.uk/scsc-156C

Assurance Case Guidance covering Challenges, Common Issues and Good Practice.
scsc.uk/scsc-159

Book Review

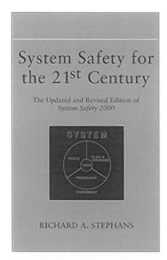

Malcolm Jones reviews the 2nd Edition of the book "System Safety for the 21st Century" by Richard A. Stephans. Richard is a co-editor of the International System Safety Society System Safety Handbook and is well known for his contributions and leadership of the Society over many decades.

Activities, processes and products are undertaken for a reason and that is for an overall benefit. However, all are subject to hazards and risk and there must be a constant process of looking for the correct benefit/risk balance – of course no activity, process product is completely free from risk – none can truly involve zero risk.

The comprehensive approach to achieving this desired balance sits under the heading of System Safety. So, what is it? It is the process and the set of support tools which enable one to minimise the hazards and detriments associated with the relevant activities. In order to accomplish this, one must define the system and its boundaries and then to both identify the hazardous conditions that can arise within the boundary together with the threats that impinge on the system from outside of the boundary, and in addition, the threats to the environment emanating from inside of the boundary.

In the former case they can arise from quality failures, including human reliability, and logic failures arising from 'as yet' not understood fault sequences. In the 'external case' they arise from failures such as poor training, environmental uncertainties and in the setting of incorrect policy and strategy. Over many decades, System Safety has evolved from a more reactive nature where improvements are made only through learning from failures – not really suitable for high consequence enterprises – to today's more pro-active form. This is now based on better fundamental understanding, better assessment processes, better standards, more comprehensive analysis tools with better audit and regulation procedures. However, unlike 'set educational subjects' such as engineering, science, technology and mathematics, there are less opportunities for formal System Safety education and training in academia and elsewhere, even though system safety impacts on all aspects of life. One hopes that this will continue to be rectified.

This leads us directly to the importance and value of this book [1], which gives a complete insight into the nature of what System Safety is all about, including its approaches, methodologies and tools, and which provides guidance on the successful application of a comprehensive, proactive approach for ensuring safe system design.

This is a book that will prove valuable to all practitioners in System Safety, ranging from the experienced proponents who need reminding of the range of procedures and techniques available for tackling the challenges that lie in front of them, to the new practitioners who are about to embark on new and fruitful careers in this exciting and valuable field – the essential guide to start them soundly on their way.

The book is written in a form whose preface directs the reader to the appropriate parts of the book to satisfy each category of interested recipient, whether safety manager, student or dedicated safety professionals. The original edition of 2004 has now been updated to include important System Safety developments that have evolved since that time and as such, brings the subject up to date.

It is not the purpose of the book to delve into detail in all specific areas, follow on detail can be found from the supporting reference list, but rather it identifies in a comprehensive fashion, the range of where and how System Safety can be applied. As such, it acts as the launching points for further detailed practitioner application on an as-needed custom basis.

Not only should System Safety be valued for its moral dimension, but a successful and well-structured safety culture is invaluable within the competitive environment which enterprises inhabit. For well-understood reasons, good safety represents a major attribute for enterprise brand and commercial success. At the extreme end lies enterprises where safety failure can be catastrophic and where the application of System Safety should be paramount. The author's valuable experience in these areas is reflected in the contents of the book.

The book also ventures into Artificial Intelligence (AI) aspects of System Safety, but this is restricted to health aspects. Of course, we are now seeing a burgeoning of AI in many other areas of System Safety, coupled with associated concerns about its probabilistic rather than deterministic relationship between cause and effect, when applied to high consequence enterprises.

The author has many decades of experience in hands-on successful application of System Safety in a wide range of areas and is a member of a cadre of pioneers in the US who established the concept of the System Safety profession, and which eventually founded what has become the International System Safety Society. He was a prominent member of that evolution and has continued to play a significant role in its subsequent developments, both in leadership and educator roles. He is a Co-Editor of the Society's "System Safety Analysis Handbook". The author was very familiar with the System Safety challenges that engineers faced in those early days and his direct involvement and experience has enabled him to clearly highlight the System Safety development history in the book. This 2nd edition brings us up to date with modern approaches.

The author has also capitalised on this experience in relation to his role as an educator, and this is, again, reflected in the style of the book and of course in its associated Instructor Manual, which forms part of this review. The Manual gives comprehensive advice on how an Educator can best develop teaching courses by way of best structuring and ordering of chapter coverage. Each chapter in turn has an associated set of questions to best support student learning through enabling a deeper and more reflective understanding of the contents of each chapter.

The author's System Safety wealth of knowledge and experience is founded in his extensive career in US Department Of Defence (DOD), Department Of Energy (DOE) and environmental restoration programmes. For this reason, the book inevitably has a strong US slant, for example with its references to US aviation, DOE, DOD, NASA, Environmental Protection Agency (EPA), Occupational Safety and Health Administration (OSHA) and the US nuclear industries. As such, its contents may not be immediately familiar to an international audience. For example, in the UK, where system safety activities are based around Relevant Good Practice, Joint Service Publications and the Ministry of Defence requirements for an enveloping Formal Safety Case, with its emphasis on demonstrating that the risk is As Low as Reasonably Practicable (ALARP). The latter being a legal requirement in the UK. Nevertheless, from a general perspective, the book's contents will be familiar, understood and applicable internationally. After all, the processes of System Safety including its problems, techniques, procedures and requirements are somewhat common the world over. For this reason, the book will not suffer from this national bias base.

The book covers the whole range of System Safety from system concept through to disposal and along the way covering all aspects of risk management, control processes, accident analysis and sound design principles. This is complemented with a comprehensive range of risk analysis tools and procedures, with examples of application given to set the reader in the right direction. Perhaps one approach that is not covered is the System-Theoretic Accident Model and Processes (STAMP) methodology advocated by Nancy Leveson of MIT.

> **System Safety practitioners within whatever areas and level of business they occupy; technical, management, medical, educational, would surely benefit from having this on their bookshelf**

This revised edition now includes a section on the value of System Safety in hospital health care and management, together with the general medical field, reminding the reader of how wide-ranging is the application of the System Safety. We are all now very well aware of its value in the field of infection control given the recent/current Coronavirus pandemic.

The book contains an extensive list of references, again mainly of a US nature, for those who require to delve in more depth into the various processes and tools of System Safety. One to note is the 1997 Edition of the System Safety Analysis Handbook, Second Edition, St Pauls MN, Published by the International System Safety Society [2].

In summary, System Safety practitioners within whatever areas and level of business they occupy; technical, management, medical, educational, would surely benefit from having this on their bookshelf and the associated Instructor Manual is a must for the latter category.

Malcolm Jones, BSc, PhD, C. Eng, C. Phys, F. int P, MBE

Malcolm is a long time Fellow of the International System Safety Society. He is a Physicist by training and has more than 50 years of experience in safety in the UK's nuclear Industry, within which he still plays an active role. During his career he has gained a number of National and International awards, including the International Systems Safety Society's development award for lifetime contributions to the development of the System Safety process.

References

[1] System Safety for the 21st Century – 2nd Edition Richard A. Stephans Published by John Wiley & Sons Inc., 2022, 111 River Street, Hoboken NJ 07030, USA. ISBN: 9781119634751 https://www.amazon.co.uk/System-Safety-21st-Century-Updated/dp/047144454

[2] System Safety Analysis Handbook, Second Edition, St Pauls MN, Published by the International System Safety Society

Training event
Clinical Data Safety
The Emerging Challenge

FACULTY OF
**CLINICAL
INFORMATICS**

NHS England

Monday 3rd April, 1 - 3.30 pm

Registration:
Via NHS: Click here
Via SCSC: scsc.uk/e1000

Friday 21st April, 1 - 3.30 pm

Registration:
Via NHS: Click here
Via SCSC: scsc.uk/e1001

Modern healthcare systems provide data that is used to make safety-critical decisions regarding patient-safety. Errors in, or incorrect use of such data, can cause harm to patients and adverse clinical outcomes. Ensuring the safe use of data is a complex challenge for the healthcare industry.

Accidents are happening...

There have been several clinical accidents and incidents where data, as distinct from purely software and hardware, has been the major cause.

The risks from data will only increase as our systems become more inter-connected, autonomous, and increasingly use data-intensive technologies, such as IoT, Artificial Intelligence, and Machine Learning.

Could your data become the cause of a clinical risk?

NHS safety standards for Clinical risk management
DCB 0129
DCB 0160

You will learn how to:

- Assess your organisation's exposure to data risks
- Reduce the risk of your data causing harm
- Understand the SCSC Data Safety Guidance and NHS Standards for Clinical risk management
- Assure the safety of your data assets
- Use the data safety analysis tool RADISH

The SCSC Data Safety Guidance

The *Safety Critical Systems Club (SCSC)* have published the "Data Safety Guidance" (scsc.uk/scsc-127H) to address this problem.

The guidance captures best practice and describes a Data Safety Management Process, which can be integrated into an overall Safety Management System.

0
101
01100

Caution
Data in use

Data Safety Guidance
Version 3.5

The Data Safety Initiative
Working Group (DSIWG)
SCSC-127H

The RADISH Tool

Risk Assessor for Data Integrity and Safety Hazards (RADISH) is a software tool from Mission Critical Applications, to aid in the application of the SCSC guidance by a practitioner.

Training Webinar delivered jointly with SCSC and MCA Ltd

MISSION CRITICAL APPLICATIONS

Connect

The Newsletter and eJournal

Do you have a topic you'd like to share with the systems safety community? Perhaps an interesting area of research or project work you've been involved in, some new developments you'd like to share, or perhaps you would simply like to express your views and opinions of current issues and events. There are now two publishing vehicles for content – shorter, more informal content, can be published in the Newsletter with longer, more technical peer-reviewed material more suitable for the eJournal. If you are interested in submitting content, then get in touch with Paul Hampton for Newsletter articles: paul.hampton@scsc.uk or John Spriggs for eJournal papers: john.spriggs@scsc.uk

The SCSC Website

Visit the Club's website thescsc.org for more details of the Safety-Critical Systems Club including past newsletters, details of how to get involved in working groups and joining information for the various forthcoming events.

Facebook

Follow the Safety-Critical Systems Club on its very own Facebook page.

www.facebook.com/SafetyClubUK

Twitter

Follow the Safety-Critical Systems Club's Twitter feed for brief updates on the club and events: @SafetyClubUK

LinkedIn

You can find the club on LinkedIn. Search for the Safety-Critical Systems Club or use the following link:

www.linkedin.com/groups/3752227

Advertising

Do you have a product, service or event you would like to advertise in the Newsletter? The SCSC Newsletter can reach out to 1,000 of individuals involved in Systems Safety and so is the perfect medium for engaging with the community. For prices and further details, please get in touch with the Newsletter Editor.

SCSC Working Groups

The Safety-Critical Systems Club is committed to supporting the activities of working groups for areas of special interest to club members. The purpose of these groups is to share industry best practice, establish suitable work and research programmes, develop industry guidance documents and influence the development of standards.

Assurance Cases

The Assurance Cases Working Group (ACWG) has been established to provide guidance on all aspects of assurance cases including construction, review and maintenance. The ACWG will:

- Be broader than safety, and will address interaction and conflict between related topics
- Address aspects such as proportionality, rationale behind the guidance, focus on risk, confidence and conformance
- Consider the role of the counter-argument and evidence and the treatment of potential bias in arguments

In Aug 2021, the group published v1.0 of the Assurance Case Guidance: scsc.uk/scsc-159

One of the working group's activities is the maintenance of the Goal Structuring Notation (GSN) Community standard.

See scsc.uk/gsn for further details.

In May 2021, the group published v3.0 of the standard: scsc.uk/scsc-141C

Lead Phil Williams phil.williams@scsc.uk

SCSC Working Groups

Security Informed Safety

The Security Informed Safety Working Group (SISWG) aims to capture cross-domain best practice to help engineers find the 'wood through the trees' with all the different security standards, their implication and integration with safety design principles to aid the design and protection of secure safety-critical systems and systems with a safety implication.

The working group aims to produce clear and current guidance on methods to design and protect safety-related and safety-critical systems in a way that reflects prevailing and emerging best practice.

The guidance will allow safety, security and other stakeholders to navigate the different security standards, understand their applicability and their integration with safety principles, and ultimately aid the design and protection of secure safety-related and safety-critical systems.

Lead **Stephen Bull** stephen.bull@scsc.uk

Data Safety Initiative

Data in safety-related systems is not sufficiently addressed in current safety management practices and standards.

It is acknowledged that data has been a contributing factor in several incidents and accidents to date, including events related to the handling of Covid-19 data. There are clear business and societal benefits, in terms of reduced harm, reduced commercial liabilities and improved business efficiencies, in investigating and addressing outstanding challenges related to safety of data.

The Data Safety Initiative Working Group (DSIWG) aims to have clear guidance on how data (as distinct from the software and hardware) should be managed in a safety-related context, which will reflect emerging best practice.

An update to the guidance (v3.5) was published in Jan 2023: scsc.uk/scsc-127H

Lead **Mike Parsons** mike.parsons@scsc.uk

SCSC Working Groups

Safety of Autonomous Systems

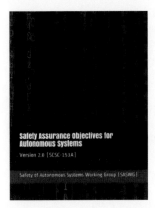

The specific safety challenges of autonomous systems and the technologies that enable autonomy are not adequately addressed by current safety management practices and standards.

It is clear that autonomous systems can introduce many new paths to accidents, and that autonomous system technologies may not be practical to analyse adequately using accepted current practice. Whilst there are differences in detail, and standards, between domains many of the underlying challenges appear similar and it is likely that common approaches to core problems will prove possible.

The Safety of Autonomous Systems Working Group (SASWG) aims to produce clear guidance on how autonomous systems and autonomy technologies should be managed in a safety-related context, in a way that reflects emerging best practice.

The group published v3 of its guidance Safety Assurance Objectives for Autonomous Systems, in Jan 2022 scsc.uk/scsc-153B

Lead **Philippa Ryan** pmrc@adelard.com

Multi- and Manycore Safety

It is becoming harder and harder to source single-core devices and there is a growing need for increased processing capability with a smaller physical footprint in all applications. Devices with multiple cores can perform many processes at once, meaning it is difficult to establish (with sufficient evidence) whether or not these processes can be relied upon for safety-related purposes.

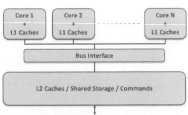

Scheduler maps processes to different cores and treats each core as a separate processor

Parallel processes need to access the same shared resources, including memory, cache and external interfaces, so they may contend for the same resources. Resource contention is a source of interference which can prevent or disrupt completion of the processes, meaning it is difficult to know with a defined uncertainty the maximum time each process will take to complete (Worst Case Execution Time, WCET) or whether the data stored in shared memory has been altered by other processes.

The Multi- and Manycore Safety Working Group (MCWG) has been established to explore the future ways of assuring the safety of multi- and manycore implementations.

Lead **Lee Jacques** Lee.Jacques@leonardocompany.com

SCSC Working Groups

Ontology

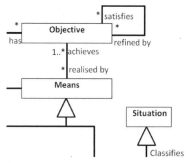

The Ontology Working Group (OWG) develops ontologies that will form the basis of SCSC guidance, as well as having wider industrial and academic applications.

The OWG is currently working on the definition of an ontology of risk for application in guidance for risk-based decision making – notably safety and security – and for which ISO 31000 Risk Management principles are to be applied.

The Data Safety Working Group (DSIWG) developed the core aspects of the Risk Ontology, which has been migrated to this working group. The Risk Ontology will form the upper ontology to the Data Safety Ontology that the DSIWG will continue to develop.

Lead **Dave Banham** ontology@scsc.uk

Service Assurance

Risks presented by safety-related services are rarely explicitly recognised or addressed in current safety management practices, guidelines and standards. It is likely that service (as distinct from system) failures have led to safety incidents and accidents, but this has not always been recognised. The Service Assurance Working Group (SAWG) has been set up to produce clear and practical guidance on how services should be managed in a safety-related context, to reflect emerging best practice.

The group published v3.1 of the guidance in Jan 2023: scsc.uk/scsc-156C

Lead **Mike Parsons** mike.parsons@scsc.uk

SCSC Working Groups

SCSC Safety Culture

The Safety Culture Working Group (SCWG) has been established to provide guidance on creating and maintaining an effective safety culture. The group seeks to improve safety culture in safety-critical organisations focussed on product and functional safety, by sharing examples and latest approaches collated from real-life case studies.

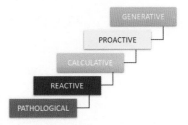

Meetings provide an opportunity to discuss any particular aspects attendees are interested in taking forward, and to help set future directions for the group.

Lead **Michael Wright** michael.wright@greenstreet.co.uk

Systems Approach to Safety of the Environment

The Systems Approach to Safety of the Environment Working Group (SASEWG) is a new group intending to apply Systems Safety practices to systems that are embedded within the natural environment, while focussing on that environment.

The group aims to produce clear guidance on how engineered systems should be developed and managed throughout their entire lifecycle so as to preserve, protect and enhance the environment.

Since holding its inaugural meeting on 17th June 2022 the group is now well underway with its 5th meeting being held in November 2022. Please get in touch with the working group lead if you would like to join, or find out more about this group.

Lead **Sarah Carrington** sarah.carrington@scsc.uk

SCSC Membership

The SCSC provides a range of services to the System Safety community including seminars, tutorials, leadership events, specialist topic working groups, the annual symposium and a comprehensive body of publications. Membership brings many valuable benefits such as free access to online events, the SCSC Newsletter and access to presentations and other resources from events.

Individual Membership

To become an individual member of the SCSC please register on the SCSC website using the ⚲ icon at the top right of any page and select "Register". Complete and save your account registration and then verify your email address. Once registered and logged in click the link "why not join the SCSC..." inviting you to become a member at the top right of the page or select "Pay membership" from the ⚲ icon.

Individual membership can be paid online using a credit/debit card through our secure payment partner Realex Global Payments or contact Alex King for other payment methods. For student or retired member rates please contact Alex King to get your account status changed.

Corporate Membership

Your company contact with the SCSC should arrange the membership and any renewals for your organisation. To join as a member covered by a corporate membership, register as per the instructions for an individual member and then contact Alex King to confirm your affiliation.

Renewing Membership

You should be notified by email when your membership is almost expired or shortly after it has expired. These notifications will contain a link to the online renewal page or you will be able to renew when logging onto the website through the 'click to renew' link.

Membership Fees

The following fees are applicable for new and renewing members:

- 1 year Individual Membership: £140
- 2 year Membership: 11% discount: £250
- 3 year Membership: 17% discount: £350 (3 years for the price of 2)
- 1 year SFI Membership: FREE for first year, £35 for years 2 & 3
- 1 year Membership, retired member rate: £35
- For Corporate Membership discounts contact Alex King.

A one-month Publication Pass is also available for £15. This allows access to all SCSC website publications in a particular calendar month.

Contact Alex King using office@scsc.uk

The SCSC Steering Group

Tom Anderson
Honorary member

Robin Bloomfield
Honorary member

Stephen Bull
stephen.bull@scsc.uk

Dewi Daniels
dewi.daniels@scsc.uk

Dai Davis
Honorary member

Jane Fenn
jane.fenn@scsc.uk

Zoe Garstang
zoe.garstang@scsc.uk

Paul Hampton
paul.hampton@scsc.uk

Louise Harney
louise.harney@scsc.uk

James Inge
james.inge@scsc.uk

Brian Jepson
brian.jepson@scsc.uk

Nikita Johnson
nikita.johnson@scsc.uk

Graham Jolliffe
Honorary member

Tim Kelly
Honorary member

Alex King
alex.king@scsc.uk

Mark Nicholson
mark.nicholson@scsc.uk

Wendy Owen
wendy.owen@scsc.uk

Mike Parsons
mike.parsons@scsc.uk

Felix Redmill
Honorary member

Roger Rivett
roger.rivett@scsc.uk

John Spriggs
john.spriggs@scsc.uk

Emma Taylor
Honorary member

Phil Williams
phil.williams@scsc.uk

Sean White
sean.white@scsc.uk

Club Positions

The current and previous (marked in italics) holders of club positions are as follows:

Managing Director

Mike Parsons 2019-

Tim Kelly 2016-2019

Tom Anderson 1991-2016

Steering Group Chair

Roger Rivett 2019-

Graham Jolliffe 2014-2019

Brian Jepson 2007-2014

Bob Malcolm 1991-2007

Programme & Events Coordinator

Mike Parsons 2014-

Chris Dale 2008-2014

Felix Redmill 1991-2008

Manager

Alex King 2019-

Honorary Solicitor

Dai Davis 2022-

Newsletter Editor

Paul Hampton 2019-

Katrina Attwood 2016-2019

Felix Redmill 1991-2016

University of York Coordinator

Mark Nicholson 2019-

eJournal Editor

John Spriggs 2021-

Administrator

Alex King 2016-

Joan Atkinson 1991-2016

Website Editor

Brian Jepson 2004-

Safety Futures Initiative Leads

Zoe Garstang 2019-
Nikita Johnson 2023-

Nikita Johnson 2019-2021

Calendar

January
M	T	W	T	F	S	S
						1
2	3	4	5	6	7	8
9	10	11	12	13	14	15
16	17	18	19	20	21	22
23	24	25	26	27	28	29
30	31					

February
M	T	W	T	F	S	S
	1	2	3	4	5
6	7	8	9	10	11	12
13	14	15	16	17	18	19
20	21	22	23	24	25	26
27	28					

March
M	T	W	T	F	S	S
		1	2	3	4	5
6	7	8	9	10	11	12
13	14	15	16	17	18	19
20	21	22	23	24	25	26
27	28	29	30	31		

April
M	T	W	T	F	S	S
					1	2
3	4	5	6	7	8	9
10	11	12	13	14	15	16
17	18	19	20	21	22	23
24	25	26	27	28	29	30

May
M	T	W	T	F	S	S
1	2	3	4	5	6	7
8	9	10	11	12	13	14
15	16	17	18	19	20	21
22	23	24	25	26	27	28
29	30	31				

June
M	T	W	T	F	S	S
			1	2	3	4
5	6	7	8	9	10	11
12	13	14	15	16	17	18
19	20	21	22	23	24	25
26	27	28	29	30		

July
M	T	W	T	F	S	S
					1	2
3	4	5	6	7	8	9
10	11	12	13	14	15	16
17	18	19	20	21	22	23
24	25	26	27	28	29	30
31						

August
M	T	W	T	F	S	S
1	2	3	4	5	6
7	8	9	10	11	12	13
14	15	16	17	18	19	20
21	22	23	24	25	26	27
28	29	30	31			

September
M	T	W	T	F	S	S
			1	2	3
4	5	6	7	8	9	10
11	12	13	14	15	16	17
18	19	20	21	22	23	24
25	26	27	28	29	30	

October
M	T	W	T	F	S	S
						1
2	3	4	5	6	7	8
9	10	11	12	13	14	15
16	17	18	19	20	21	22
23	24	25	26	27	28	29
31						

November
M	T	W	T	F	S	S
	1	2	3	4	5
6	7	8	9	10	11	12
13	14	15	16	17	18	19
20	21	22	23	24	25	26
27	28	29	30			

December
M	T	W	T	F	S	S
			1	2	3
4	5	6	7	8	9	10
11	12	13	14	15	16	17
18	19	20	21	22	23	24
25	26	27	28	29	30	31

Events Diary

7-9 Feb 2023
SCSC Symposium

**Safety Critical
Systems Symposium
SSS'23**

York, UK and online

scsc.uk/e898

3 Apr 2023
SCSC Webinar

**Clinical Data Safety:
The Emerging
Challenge**

**1pm-3.30pm
online**

scsc.uk/e1000

6 Apr 2023
SCSC Seminar

**Safety of Autonomy
in Complex
Environments**

**London, UK and
blended online**

scsc.uk/e890

21 Apr 2023
SCSC Webinar

**Clinical Data Safety:
The Emerging
Challenge**

**1pm-3.30pm
online**

scsc.uk/e1001

8 June 2023
SCSC Seminar

**How to Write
Compelling Safety
Arguments**

London, UK

scsc.uk/e982

3-8 Sep 2023
Conference

ESREL 2023: European
Conference on Safety and
Reliability (ESREL)

**University of
Southampton, UK**

esrel2023.com

Printed in Great Britain
by Amazon

17626210R00031